Jeetkumar Mehta
Yiqiang Yu
Zhizhang Chen

Wireless Power Transfer Design For Small Implantable Medical Devices

Jeetkumar Mehta
Yiqiang Yu
Zhizhang Chen

Wireless Power Transfer Design For Small Implantable Medical Devices

Via use of Magnetic Resonance Coupling

LAP LAMBERT Academic Publishing

Impressum / Imprint
Bibliografische Information der Deutschen Nationalbibliothek: Die Deutsche Nationalbibliothek verzeichnet diese Publikation in der Deutschen Nationalbibliografie; detaillierte bibliografische Daten sind im Internet über http://dnb.d-nb.de abrufbar.
Alle in diesem Buch genannten Marken und Produktnamen unterliegen warenzeichen-, marken- oder patentrechtlichem Schutz bzw. sind Warenzeichen oder eingetragene Warenzeichen der jeweiligen Inhaber. Die Wiedergabe von Marken, Produktnamen, Gebrauchsnamen, Handelsnamen, Warenbezeichnungen u.s.w. in diesem Werk berechtigt auch ohne besondere Kennzeichnung nicht zu der Annahme, dass solche Namen im Sinne der Warenzeichen- und Markenschutzgesetzgebung als frei zu betrachten wären und daher von jedermann benutzt werden dürften.

Bibliographic information published by the Deutsche Nationalbibliothek: The Deutsche Nationalbibliothek lists this publication in the Deutsche Nationalbibliografie; detailed bibliographic data are available in the Internet at http://dnb.d-nb.de.
Any brand names and product names mentioned in this book are subject to trademark, brand or patent protection and are trademarks or registered trademarks of their respective holders. The use of brand names, product names, common names, trade names, product descriptions etc. even without a particular marking in this work is in no way to be construed to mean that such names may be regarded as unrestricted in respect of trademark and brand protection legislation and could thus be used by anyone.

Coverbild / Cover image: www.ingimage.com

Verlag / Publisher:
LAP LAMBERT Academic Publishing
ist ein Imprint der / is a trademark of
OmniScriptum GmbH & Co. KG
Heinrich-Böcking-Str. 6-8, 66121 Saarbrücken, Deutschland / Germany
Email: info@lap-publishing.com

Herstellung: siehe letzte Seite /
Printed at: see last page
ISBN: 978-3-659-34590-6

Zugl. / Approved by: Halifax, Dalhousie University., 2013

Copyright © 2015 OmniScriptum GmbH & Co. KG
Alle Rechte vorbehalten. / All rights reserved. Saarbrücken 2015

Table of Contents

Chapter 1 Introduction 1
  1.1   Background of WPT Technology 1
  1.2   Types of Wireless Power Transmission: 2
      1.2.1   Near field techniques 2
          1.2.1.1 Inductive Coupling 2
          1.2.1.2 Magnetic Resonance Coupling 3
      1.2.2   Far-Field Energy Transfer Techniques 4
          1.2.2.1 Microwave Power Transmission (MPT) 5
          1.2.2.2 LASER Technology 5
  1.3   Selection of WPT Techniques for charging Implantable Medical Device 6
  1.4   Motivation 7
  1.5   Objective 7
  1.6   Organization of this Report 8

Chapter 2 Literature Review 9
  2.1   Brief History Wireless Power Transmission 9
  2.2   The state – of the Art on WPT 11
      2.2.1   Basic WPT model using Magnetic Resonance Coupling 11
      2.2.2   WPT for Smaller Devices 12
      2.2.3   WPT in Medical Implants 12
  2.3   Types of Implantable Medical Devices (IMD) 15
  2.4   Infeasibility of WPT in Inductive Coupling in Charging Pacemaker… 16

Chapter 3 Design Theory and Proposed WPT System 18

i

- 3.1 Operational principle of Magnetic Resonance Coupling 18
- 3.2 Design Theory for Strong Coupling in Magnetic Resonance Coupling Using CMT 18
- 3.3 Equivalent Circuit Model 22
- 3.4 The proposed WPT System 25

Chapter 4 Simulation, Fabrication and Testing setup of the Design 27

- 4.1 Simulation Results 27
  - 4.1.1 Design #1 27
  - 4.1.2 Design #2 29
  - 4.1.3 Design #3 30
- 4.2 Prototype 31
- 4.3 Testing Set-Up 33

Chapter 5 Measurements and Discussions 36

- 5.1 Design #2 36
- 5.2 Design #3 39
- 5.3 Performance Comparison of Three Proposed Designs at 10 cm Transmission Distance……………………………………………..45
- 5.4 Comparison of Final Proposed Design #3 with Fei Zhang Design...46

Chapter 6 Conclusion and Future work 47

- 6.1 Conclusion 47
- 6.2 Future Work 47

Bibliography……………………………………………………..48

List of Figures

Figure 1.1 Inductive coupling technique 2

Figure 1.2 Magnetic resonance coupling [2] 3

Figure 1.3 Tesla wardenclyffe tower [5] 4

Figure 1.4 Space based Micro Wave Power transmission [7] 5

Figure 1.5 LaserMotive's receiver visible mounted under the wing of Lockheed Martin's Stalker unmanned aircraft during laser-powered test flights in the summer of 2012 [8] 6

Figure 2.1 Faraday's Law Example... 10

Figure 2.2 Experimental and theoretical values of received power with distance are compared (a) Transferred efficiency over distance (b) Received power over distance [10]. 11

Figure 2.3 System demonstration for medical implant (a) through the air (b) through the head model (c) agar solution and internal view of head model (d) receiver implanted in "brain" and the agar cover [12]. 13

Figure 2.4 Experimental set up on pig a) implantable receiver b) power transfer to implanted receiver [12]... 14

Figure 2.5 Types of Implantable Medical Devices [13] 16

Figure 2.6 Inductively coupling method of charging pacemaker [14]........ 17

Figure 3.1 Schematic of the wireless power transfer system using magnetic coupling resonator technique. 23

Figure 3.2 Equivalent circuit of the WPT system. 23

Figure 3.3 The proposed WPT system using magnetic resonance coupling. 26

Figure 4.1 Simulated scattering parameters of Design #1 28

Figure 4.2 Effeciency vs transmission distance for Design #1 29

Figure 4.3 Simulation s-parameters for Design #2 29

Figure 4.4 Simulated scattering parameters at 8 cm transmission distance for Design #3.. 30

Figure 4.5 Receiver coils of 100 mm in diameter 32

Figure 4.6 Receiver coils of 50 mm in diameter 32

Figure 4.7 Test set up diagram 33

Figure 4.8 Experimental set up 34

Fig 4.9 Agilent VNA(Vector Network Analyzer) E5071B 34

Figure 4.10 Agilent Vector Signal Generator E4438C 35

Figure 5.1 Measurement results for Design #2 37

Figure 5.2 Efficiency vs distance for simulation and measurement for Design #2 38

Figure 5.3 Experimental set up without pork for Design #3 40

Figure 5.4 Experimental set up with pork for Design #3 40

Figure 5.5 Measurement results at 8 cm transmission distance for Design #3 41

Figure 5.6 Efficiency vs Distance for measurement (with/without pork) for Design #3 42

Figure 5.7 Orientation angles of receiver coils for Design #3 43

Figure 5.8 Efficiency vs Orientation angle for Design #3 44

List of Tables

Table 1: Parameters of Coils for proposed Three Designs 26

Table 2: Efficiency (simulation/measurement) vs Distance for Design #2. 38

Table 3: Efficiency(Simulation/Measurement) vs transmission distance [with/without Pork] for Design #3 42

Table 4: Scattering Parameters S21 at 10 cm Transmission Distance for Three proposed Designs 45

Table 5: Comparison with Fei Zhanf Design [12] 46

Abstract

Efficient and compact wireless power transfer (WPT) systems are proposed and designed for recharging small implantable medical devices. They use magnetic resonance coupling scheme to transfer power over a relatively large distance. The receiver resonator coil and the load loop are designed in correspondence to size restrictions of implantable devices. The dimensions of the coils are optimized and effective values of the lumped capacitors are investigated and fine-tuned for efficiency enhancement. Three design configurations of the WPT system, each consisting of two coils for the transmitting side and two coils for the receiving side, are designed and fabricated. The transfer efficiency is measured with different transmission distances and with different orientation angles of the receiving coils. The measurement results show good agreements with the simulations and illustrate that the proposed WPT systems exhibit nearly omnidirectional performance. Furthermore, the receiver coils are implanted inside of a biological object to show the power can be transferred effectively.

List of Abbreviations

WPT	Wireless Power Transmission
IMDs	Implantable Medical Devices
AC	Alternating Current
MIT	Massachusetts Institute of Technology
MPT	Microwave Power Transmission
SPS	Solar Power Satellite
DC	Direct Current
EM	Electromagnetic
MHz	Mega Hertz
LED	Light Emitting Diode
IACUC	Institutional Animal Care and Use Committee
CMT	Couple Mode Theory
RLC	Resistor Inductor Capacitor
TX	Transmitter
RX	Receiver
Pf	Pico Faraday
HFSS	High Frequency Structural Simulator
VNA	Vector Network Analyzer
VSG	Vector Signal Generator
SMA	SurfaceMiniature version A
SAR	Specific Absorption Rate
RFID	Radio Frequency Identification

Acknowledgements

I dedicate this project to my family. Their love, trust and support have given me great comforts and encouragements throughout my student life. I can never thank them enough.

I am extremely thankful to my supervisor Dr. Zhizhang (David) Chen for providing me with immense support, help and guidance throughout my Master's. I would like to specially thank Dr. Yiqiang(John) Yu and Farid Jolani for their kind support throughout my project work. I won't be able to achieve what I have achieved today without my group members.

I would like to devote this work to my parents Asha & Atul, and my beloved brother Arjun.

Chapter 1   Introduction

This chapter deals with the general introduction and the basic idea about wireless power transmission (WPT) technology. Section 1.1 and 1.2 provides the background of WPT technology and types of WPT respectively. Section 1.3 represents remarks on various WPT techniques. Section 1.4 describes the motivation behind this project followed by the main objectives in Section 1.5. Section 1.6 contains a brief outline of the report.

1.1   Background of WPT Technology

Wireless Power Transmission (WPT) is a phenomenon in which electrical energy is transmitted wirelessly from a power source to a load without any physical connection. This technology is extremely useful when connecting a wire with a power source is difficult, nearly impossible or hazardous. Due to its simplicity, WPT opens a vast area for developing various applications which can be utilized in our day to day lives.

Wireless Power Transmission has been researched for almost 11 decades, but studies in WPT have slowed down because of technical aspects of bottle necks and lack of demands [1]. The increase in the use of electronic gadgets in the past decade has revived the interest of researchers to find a way to wirelessly charge phones, laptops, Bluetooth devices, light bulbs and IMDs (Implantable Medical Devices).

Today, even the most advanced wireless devices need a power cord to recharge their batteries. Eventually a time of demands will come when all electronic devices will be charged wirelessly.

1.2 Types of Wireless Power Transmission:

Wireless power transmission can be categorized into two categories:

- Near field
- Far field

1.2.1 Near field techniques

1.2.1.1 Inductive Coupling

Two devices are said to be mutually inductively coupled or magnetically coupled when they are configured such that change in current through one wire induces a voltage across the ends of the other wire by electromagnetic induction. This is due to the mutual inductance as shown in Figure 1.1. Transformer is an example of inductive coupling.

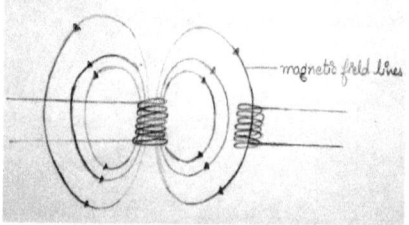

Figure 1.1 Inductive coupling technique

1.2.1.2   Magnetic Resonance Coupling

Magnetic coupling occurs when two objects exchange energy through their varying magnetic or oscillating fields. Resonance coupling occurs when the natural frequencies of both objects are the same.

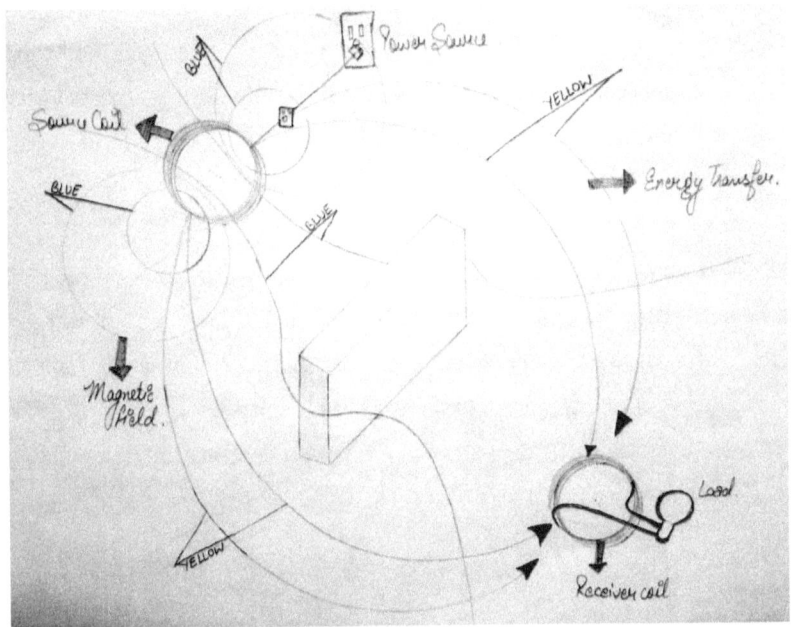

Figure 1.2 Magnetic resonance coupling [2]

Figure 1.2 shows magnetic resonance coupling where the source is connected to AC power source. Blue lines in the figure represent the induced near magnetic field generated by the coil that is connected to the AC power. The yellow lines represent the flow of energy from the source coil to the coil which is loaded with the bulb. Magnetic field represented by blue line also indicates how it wraps around conductive obstacles between coils and capturing device. The MIT research group led by Prof. Soljacic used the concept of magnetic resonance coupling to light the bulb at a distance of 2 meters [3].

a) Comparison between Magnetic resonance coupling and Inductive Coupling

By using the magnetic resonance coupling the need to be in close proximity or being in contact is eliminated as magnetic resonance coupling has greater range than inductive coupling. In addition, Magnetic resonance coupling can be one- to- many whereas inductive coupling is one-to-one.

1.2.2 Far-Field Energy Transfer Techniques

Far Field Energy Transfer technique is mainly dependent on radiative techniques. Here wave are either broadcasted in the form of narrow beam transmission of radio, or light waves that is solely for high power transfer. Tesla already gave the concept to the world on his paper: "Truly Wireless" in 1890. He constructed large Wardenclyffe Tower as shown in Figure 1.3, to transfer the energy over large distances [4].

Figure 1.3 Tesla wardenclyffe tower [5]

There are basically two methods so far for long distance WPT: the microwave power transmission and the power transmission using LASER.

1.2.2.1 Microwave Power Transmission (MPT)

MPT involves conversion of energy into microwave and then transfers the microwave through highly directional antenna (arrays) from the transmitter to the receiver through the rectenna (rectifier and antenna) which will convert microwaves into the conventional electrical power [6]. Figure 1.4 shows an example of MPT.

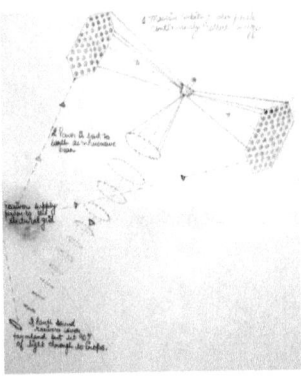

Figure 1.4 Space based Micro Wave Power transmission [7]

1.2.2.2 LASER Technology

Another possible way of wireless power transmission is to use LASER as the medium that carries electric power. The terminology is similar to microwave power transmission, but the energy emission is of high frequency and is coherent. Research organizations like NASA, ENTECH, and UAH have been working on this scheme as one of the means to transmit power wirelessly [3]. An advantage of LASER power transmission is high aperture collection efficiency which enables the antennas used for the transmission to be of small size.

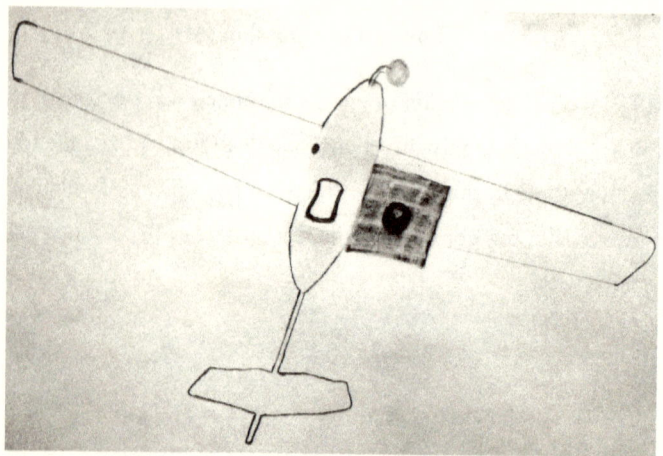

Figure 1.5 LaserMotive's receiver visible mounted under the wing of Lockheed Martin's Stalker unmanned aircraft during laser-powered test flights in the summer of 2012 [8]

LASER transmission does not get dispersed for long distance but it is attenuated when it propagates though the atmosphere. During the design, the receiver used can be simple like photovoltaic cell. It is simpler and more cost efficient than microwave power transmission. Figure 1.5 shows the application of laser technology in test flight.

1.3 Selection of WPT Techniques for charging Implantable Medical Device

Based on above descriptions, inductive coupling and magnetic resonance coupling are found to be more suitable for the design of IMDs, since MPT and Laser techniques are of radiative and high power density in nature and may cause harm to human body.

Both inductive coupling and magnetic resonance coupling are not affected much by biological bodies as energy is transmitted via magnetic field. The magnetic resonance technique is more appropriate for our design purposes than inductive coupling technique with the same dimensions; it provides larger transmission distance with little sacrifice in transmission efficiency. This is a more desired feature for deeply-implantable medical devices.

1.4 Motivation

The motivations for the present research work are as follow:

- To design a system based on magnetic resonance coupling which has transmission distance of 10 cm; the previous reported distances are all less than 10 cm for IMD.
- To improve the transmission distance between the transmitter and the IMD receiver without compromising on efficiency.
- To make the system perform evenly in all directions.
- To reduce the system size for suitable applications in IMD.

1.5 Objective

The major objective of this project is to design a WPT system that can be utilized for charging IMDs wirelessly.

The following objectives are achieved during the design process:

- Study of the existing methods for WPT and understand the requirements and parameters needed to build a robust WPT system.
- Determination of an appropriate WPT technique that is not harmful for humans when exposed and suitable for design purposes.

- Development of a WPT system for IMDs that can transmit power up to a distance of 10 cm with sufficient efficiency with the aid of EM modeling and simulation tools.
- Fine tuning and optimization of the design via analysis of differences between measurement results and the simulation results.

1.6 Organization of this Report

The remainder of the project is organized as follow:

Chapter 2 presents the history, existing and proposed designs of WPT using magnetic resonance coupling technique followed by current implantable medical devices in the market. Couple Mode theory and our three proposed designs with their dimensions are presented in chapter 3. Chapter 4 discusses the simulation, fabrication and testing. Further, chapter 5 deals with the evaluation of the design through lab testing. Chapter 6 concludes with the main contribution of the present research. This chapter also gives some directions for future research in this area.

Chapter 2   Literature Review

This chapter presents history of WPT and recent work done on WPT using magnetic resonance coupling. There have been a number of WPT designs proposed for transmitting power wirelessly for various applications using magnetic resonance coupling in the past. It is vital to study these approaches in order to understand their pros and cons, and thus help in understanding the fundamental concepts of designing a WPT structure for our purpose.

2.1   Brief History Wireless Power Transmission

The WPT concept can be tracked back to 1820, when Andre-Marie Ampere proposed a law which states that an electric current produces a magnetic field. Inspired by the following work by Michael Faraday (1830), James Maxwell (1864) and Heinrich Hertz (1888), Nikola Tesla experimentally showed transfer of wireless energy in 1891 [9]. In Tesla's experiment, he designed a resonant circuit that coupled electric current in another resonant circuit of the same structure, and with them he was able to light up a bulb wirelessly.

The contribution of Tesla was that he developed a circuit that can generate and receive a power through a time-varying magnetic field in free space. Tesla's method is a near field phenomenon whereas the commonly known propagation of an electromagnetic wave is a far field phenomenon. The two phenomenon differ by the transmission distance and angular coverage of the power transmission. Even though the near field transmission is of short range, field is more confined in space than the far field transmission.

The theory of wireless power transfer is already explained in Maxwell's equations:

$$\oiint \vec{D}.d\vec{s} = Q \tag{2.1}$$

$$\oiint \vec{B}.d\vec{s} = 0 \tag{2.2}$$

$$\oint \vec{E}.d\vec{l} = -\iint \frac{\partial \vec{B}}{\partial t}.d\vec{s} \tag{2.3}$$

$$\oint |\vec{E}|.d\vec{l} = \iint \frac{\partial \vec{D}}{\partial t}.d\vec{s} + 1 \tag{2.4}$$

The two equations 2.3 and 2.4 describe how a time-varying magnetic flux generates an electric field, and how a time-varying electric flux generates the magnetic field respectively as shown in Figure 2.1. In other words if a time varying current is generated, this time varying current will induce time varying magnetic field. This time varying magnetic field in turn can be used to generate an electric field or AC current across the receiving load.

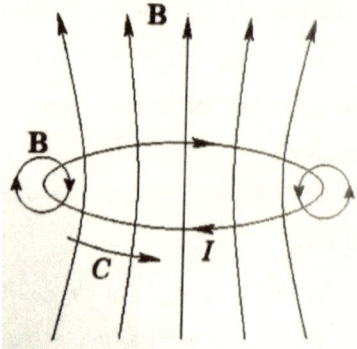

Figure 2.1 Faraday's Law Example

The following section gives insight to the work done in WPT using magnetic resonance coupling technique that forms the base for our report work.

2.2 The state – of the Art on WPT

2.2.1 Basic WPT model using Magnetic Resonance Coupling

The research team at MIT led Marin Soljacic in 2007 reported the transfer of non-radiative power to light a 60 watts bulb at a distance of 2m with an efficiency of ~40% using magnetic resonance coupling [10].

In their experimental setup, they used two identical helical coils of copper; one was for source and the other for load for magnetic coupling. They used copper wires with height $h = 20$ cm, cross-sectional radius of copper wire $a = 3$mm, radius $r = 25$ cm and number of turns $n = 5.25$. The resonant frequency at which this experiment was carried out was $f_0 = 9.56 \pm 0.3$ MHz

The comparisons of theoretical and experimental values in terms of efficiency versus distance are shown in Figure 2.2.

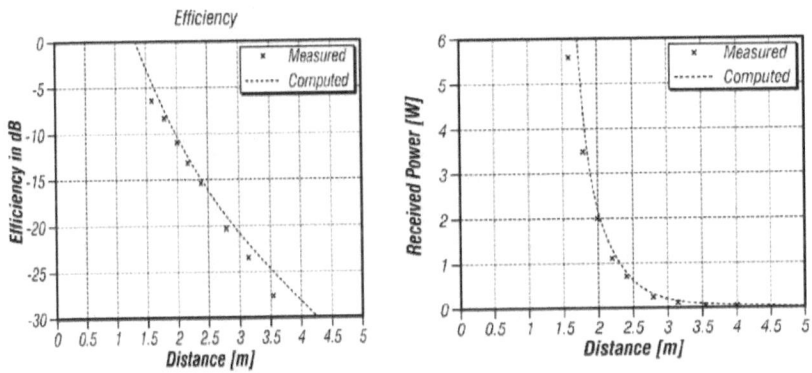

Figure 2.2 Experimental and theoretical values of received power with distance are compared (a) Transferred efficiency over distance (b) Received power over distance [10].

However, the dimensions of the receiving coils they used were too large to be considered for charging the IMDs. Thus, considerably smaller receiver coils needs to be designed to serve our purpose to charge IMDs.

2.2.2 WPT for Smaller Devices

J.Choi et al. [11] have used an efficient wireless power transmission method of magnetic resonance coupling for small devices. They used large transmitting coils and small receiving coils. Therefore, this model is popular in charging small electronic devices.

In their experimental setup for charging an iPhone 4, the receiving structure designed was of the same dimension as that of the iPhone 4. The transmitting coils used were of a circular shape with a driver coil of 40 cm and transmitter coupling coil of 60 cm in diameter, with a gap of 3 cm between them. The receiving side consisted of rectangular shaped coils, similar to the dimensions of an iPhone 4, with a gap of 0.5 cm between them. Efficiency of 75% and 40 % was achieved at a distance of 20 cm and 50 cm respectively. The resonance frequency was 13.56 MHz [11].

Although, the efficiency was sufficient to charge IMDs, the dimensions of the receiver coils were still too large to be considered to be integrated into an IMD. Hence, specific research on designing compact receivers is still needed which is the objective of this report.

2.2.3 WPT in Medical Implants

The research group of Fei Zhang et al [12] at the University of Pittsburgh designed a frequency adjustable magnetic resonance coupling system to power implantable devices and medical sensor networks. In their work, they used the concept of magnetic resonance coupling, based on which they designed and fabricated a WPT system and gave a practical demonstration for a medical application. In their practical demonstration, they used LED (Light Emitting Diode) as a load at the receiver side.

Their transmitter consisted of a driving coil and a copper tube of 165 mm in diameter. The receiver consisted of thin silver coil which was 41mm in diameter and a LED as a load.

To examine the feasibility of the receiver which they designed, they implanted the receiver inside the plastic skull model as shown in figure. 2.3 (a) and (b). The inside of this skull is filled with an agar solution (16 grams of dry agar, 0.9 liter of water and 0.6 gram of Nacl) as shown in the Fig 2.3 (c and d).

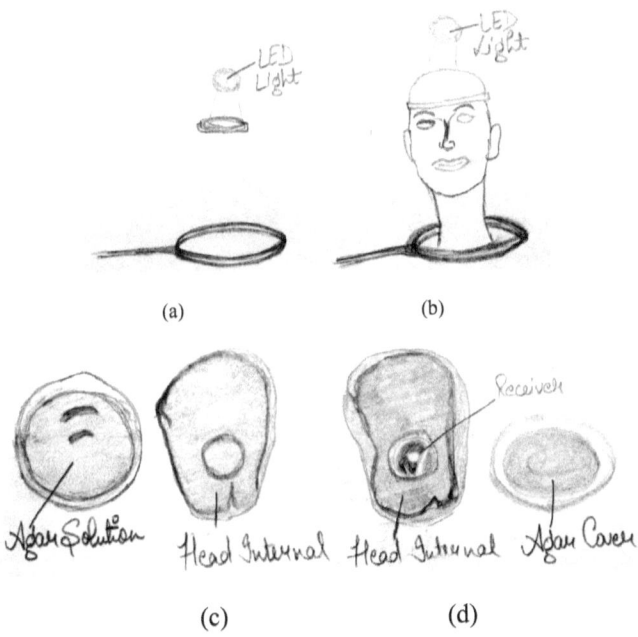

Figure 2.3 System demonstration for medical implant (a) through the air (b) through the head model (c) agar solution and internal view of head model (d) receiver implanted in "brain" and the agar cover [12].

With this experimental setup they were able to transfer power with an efficiency of 22.23% at a transmission distance of 90 mm between the transmitter and the receiver [12].

After achieving the aforementioned results, they obtained approval from IACUC (Institutional Animal Care and Use Committee) and undertook an experiment on a Yorkshire pig at the Animal Research Facility of University of Pittsburgh. They placed a receiver enclosed in plastic 35 mm beneath the pig's abdomen with LED connected to it. After all the skin was sealed up, LED was lightened up from a distance of 100 mm as seen in the Figure 2.4.

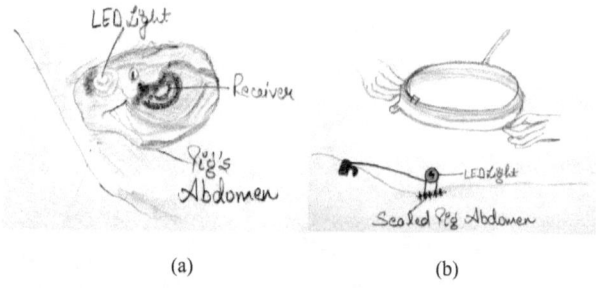

(a)   (b)

Figure 2.4 Experimental set up on pig a) implantable receiver b) power transfer to implanted receiver [12]

In conclusion to this experiment, the authors commented that if the LED is replaced by a rectifier and a regulation circuit, the power produced can be stored or can be used to charge implantable medical devices [12].

In the above design, though the receiver coils were made small enough to be utilized for IMDs, they still had to compromise on efficiency as the efficiency obtained was 22.23% at 90 mm transmission distance. Thus, a systematic WPT system needs to be developed with a higher efficiency.

2.3 Types of Implantable Medical Devices (IMD)

The types of commercially available IMD's are shown as follow:

(a) Implantable hearing device

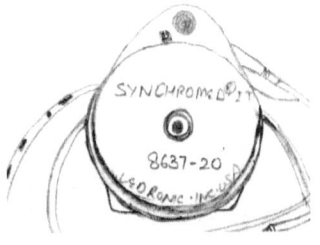

(b) Implantable drug infusion pump

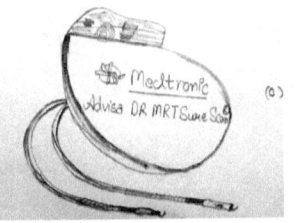

(c) Implantable pacemaker

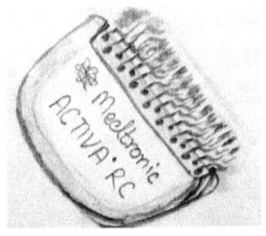

(d) Implantable neuro-stimulator

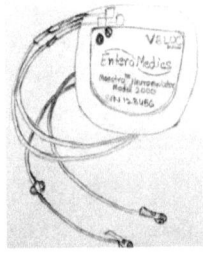

(e) Implantable vagus nerve stimulation system to cure obesity

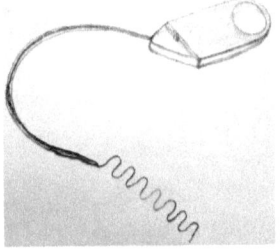

(f) Implantable Bone growth stimulator

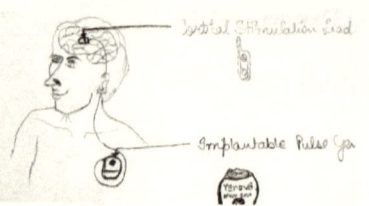

(g) Implantable Cortical stimulation system

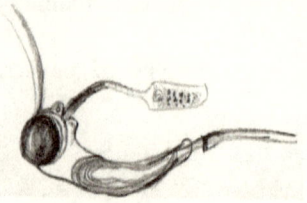

(h) Implantable Retinal prosthesis

(i) Implantable deep brain stimulation system

(j) Implantable Baroflex stimulator for the treatment of hypertension

Figure 2.5 Types of Implantable Medical Devices [13]

2.4 Infeasibility of WPT in Inductive Coupling in Charging Pacemaker

If an inductive coupling is used in charging implanted pacemaker, the pacemaker placed inside a human body needs to have small coil enclosed in its housing; an external coil then required to be placed right on human chest as shown in Fig 2.6. The two coils can inductively couple to one another as they are co-linear. The coil is placed in direct contact with human body [14]; therefore, it is not very convenient and also requires right positioning.

Figure 2.6 Inductively coupling method of charging pacemaker [14]

However, our proposed design use magnetic resonance coupling which will not only eliminate the direct contact to human body, but will also provide better efficiency and transmission distance.

Chapter 3   Design Theory and Proposed WPT System

This chapter present design theory and system for strong coupling using magnetic resonance coupling in WPT, and the equivalent circuit model. They are followed by our three proposed WPT systems with different coil dimensions and lumped elements.

3.1   Operational principle of Magnetic Resonance Coupling

In order to understand the concept of magnetic resonance coupling, assume two copper coils of the same diameter and size resonating at the same frequency. One coil is connected to the power source (transmitter) and the other coil is connected to the load (receiver). The electric power connected to the transmitter coil causes it to oscillate at a particular frequency. As a result of this the non-radiative magnetic field is generated around the transmitter coil. This magnetic field transfers power to the receiver coils and causes it to oscillate at the same frequency. This process is called 'coupled resonance' and this is the operational principle of magnetic resonance coupling.

In order to achieve high efficiency using magnetic resonance, strong coupling between individual coils plays a vital role. Couple Mode Theory (CMT) [15], [16] gives a simple and accurate way of designing the system with proper energy transmission efficient.

3.2   Design Theory for Strong Coupling in Magnetic Resonance Coupling Using CMT

The principles of Couple Mode Theory to create a system of strongly coupled magnetic resonating fields between two inductor- capacitor circuits has been demonstrated in this section.

In order to understand Haus's CMT a simple, lossless LC circuit needs to be analyzed. A suitable LC circuit used creates an oscillating magnetic field is the Colpitts Oscillator [17]. This particular circuit is desirable because of its ability to adjust circuit's capacitance so that circuit oscillates at its resonant frequency. Capacitive and inductive reactance in an electrical system reduces the amount of current flow [18] and adding resistance within the circuit. The capacitive reactance of the circuit is given by $X_C = \frac{1}{\omega C}$, where ω = circuit frequency; C = capacitance. The inductive reactance is given by $X_l = \omega L$ where L = inductance. The resonant frequency is achieved when both capacitive and inductive reactance of a circuit is equal.

$$\omega L = \frac{1}{\omega C} \quad (3.1)$$

Therefore, the resonant frequency, ω_0, is:

$$\omega_0 = \frac{1}{\sqrt{LC}} \quad (3.2)$$

Substituting the net capacitance into above equation, the natural resonating frequency of an oscillating, lossless Colpitts LC circuit is:

$$\omega_0 = \frac{1}{\sqrt{L\left(\frac{c_1 c_2}{c_1 + c_2}\right)}} \quad (3.3)$$

Lossless LC circuits are only considered for this analysis, as circuit's resistances are minimal and energy is to be conserved. The addition of circuit resistances and losses due to radiation and absorption of energy contribute to the intrinsic decay rate, Γ_m, of a circuit, which altar's objects resonant frequency as $\omega_0 = \beta - \Gamma$, where β is the resonant frequency without losses. $\Gamma = 0$, as LC circuit is considered to be lossless.

Moving on to Couple Mode Theory, CMT states that when normal modes, which are the wave patterns that moves sinusoidally at the same phase and frequency, couple to one another. The energy transferred between objects will be far greater than without coupling conditions. The system of equations in CMT can be properly understood by reviewing how Haus used simple, lossless LC circuit to derive equations of amplitudes $a_{1,\,2}$. Haus begins his description of CMT by defining the voltage and current within an LC circuit to be:

$$v = L\frac{di}{dt} \tag{3.4}$$

$$i = -C\frac{dv}{dt} \tag{3.5}$$

On combining above two equations, the second order differential equation for voltage is derived, which is same as the one for simple harmonic motion.

$$\frac{d^2v}{dt^2} + \omega_0^2 v = 0 \tag{3.6}$$

Where, ω_0 is the resonant frequency.

The above differential equation is solved assuming that voltage has exponential time dependence where $\vartheta(t) = e^{\lambda t}$. By substitution, the eigenvalue becomes, $\lambda = \pm j\omega_0^2$. This simplifies to:

$$\vartheta(t) = |V|\cos(\omega_0 t + \emptyset) \tag{3.7}$$

Where $\emptyset$ = phase, ω_0 = resonant frequency.

Similar approach can be taken to solve equations (3.4) and (3.5) for current, which results in:

$$i(t) = \sqrt{\frac{C}{L}}\,|V|\sin(\omega_0 t + \emptyset) \tag{3.8}$$

Using the definitions from equations (3.7) and (3.8), Haus then makes an interesting definition of the complex variable $a_\pm$ so that the energy within LC circuits 1 or 2 can be described.

$$a_\pm = \sqrt{\frac{C}{2}}\left(v + j\sqrt{\frac{L}{C}}\,i\right) \qquad (3.9)$$

Where $j = \sqrt{-1}$.

Substituting equation (3.7) and (3.8) in equation (3.9), this becomes:

$$a_\pm = \sqrt{\frac{C}{2}}\, V e^{j\omega_0 t} \qquad (3.10)$$

From this definition, Haus is able to display the mode amplitude $a_\pm$ exponential time dependence. In addition to this, Haus's definition of a_+ can be used to model the energy within LC circuit, E, by noting:

$$|a_+|^2 = \frac{C}{2}|V|^2 = E \qquad (3.11)$$

The coupling of two resonant modes can be further explained using Couple Mode Theory. In CMT, each mode consists of forward and backward wave. The amplitudes of these waves resonate at system's natural frequency, when modes are coupled. This motion is very efficient for passing and receiving of energy between two objects. The amplitudes between two uncoupled oscillators is found by taking derivative of equation (3.4) with respect to time and then using proper substitution equations (3.4) and (3.5):

$$\frac{da_1}{dt} = \pm j\omega_1 a_1 \qquad (3.12)$$

$$\frac{da_2}{dt} = \pm j\omega_2 a_2 \qquad (3.13)$$

Where, ω_1, ω_2 are natural resonant frequencies of each object, $j = \sqrt{-1}$ and $a_{1,2}$ = mode amplitudes.

If the amplitudes $a_{1,2}$ of two wave are able to couple by some constant k_{12}, then energy within LC circuit can be modeled using resonant energy exchange system known as CMT. The mode amplitudes, and thus the circuit energy, are modeled [16] by:

$$\frac{da_1}{dt} = j\omega_1 a_1 + k_{12} a_2 \qquad (3.14)$$

$$\frac{da_2}{dt} = j\omega_2 a_2 + k_{21} a_1 \qquad (3.15)$$

Where, k_{12} and k_{21} are the coupling coefficients between the two systems and $|a_{1,2}|^2$ = Energy within circuit 1 or 2.

The above system of equations has been used by Kurs, Karalis [19]. The system analyzed here is the simplification of karali's model that assumes two circuits are lossless and undriven, oscillating LC circuits. The purpose of this simplification is to show how CMT is used to increase the transfer of energy between two inductive circuits.

3.3 Equivalent Circuit Model

The proposed wireless power transfer (WPT) system utilizes the magnetic resonance coupling. The key elements for designing such a system are the compensation capacitors that are connected to each coil and used to achieve resonances between the large transmitter resonator coils and the small receiver resonator coils as discussed earlier.

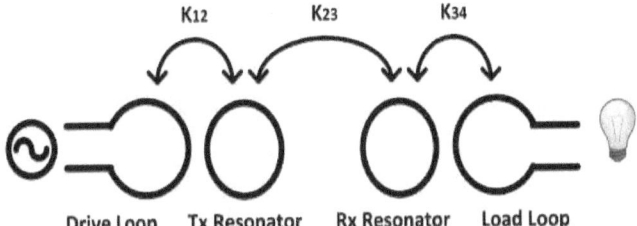

Figure 3.1 Schematic of the wireless power transfer system using magnetic coupling resonator technique.

K_{ij} is the coupling coefficient between coil i and coil j.

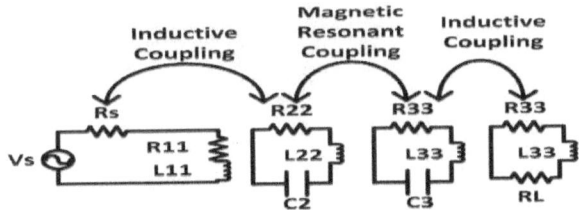

Figure 3.2 Equivalent circuit of the WPT system.

The equivalent circuit of the system depicted in Figure 3.1 is shown in Figure 3.2. The four coils are modeled as a RLC network with C_2 and C_3 being the compensation capacitors. Theoretically, all four coils are coupled among each other, but the cross couplings beyond the nearby coils are very weak and thus can be neglected.

In the equivalent RLC circuit, the self-inductance of a loop is calculated as [20]

$$L = \mu_0 a \left(\ln(\frac{8a}{r}) - 2 \right) \qquad (3.16)$$

where μ_0 is the space permeability, 'a' is the radius of the loop and 'r' is the radius of the wire.

For perfectly aligned coils, the mutual inductance of two parallel single-turn coils can be calculated as

$$M(a, b, \rho = 0, d) = \mu_0 \sqrt{ab}\left[\left(\frac{2}{k} - k\right)K(k) - \frac{2}{k}E(k)\right] \quad (3.17)$$

where,

$$k = \left(\frac{4ab}{(a+b)^2 + d^2}\right)^{1/2} \quad (3.18)$$

and $K(k)$ and $E(k)$ are the complete elliptical integrals of the first and second kind, respectively [20].

Considering the skin effect, the loss resistance of a loop is

$$R_L = \frac{\pi a}{r}\left(\frac{f\mu_0}{\pi\sigma}\right) \quad (3.19)$$

where σ is the conductivity of the wire and f is the operating frequency.

To achieve identical resonant frequencies for the transmitter and receivers coils, the lumped capacitors (C_{Tx}, C_{Rx}) shown in Fig. 3.3 can be calculated as follows:

$$C = \frac{1}{(2\pi f)^2 L} \quad (3.20)$$

The imaginary part of the equivalent impedance of a WPT system using magnetic resonance coupling can be assumed to be zero while it operates in resonance. Therefore, the power transfer efficiency can be calculated as

$$\eta = \frac{output\ power}{input\ power} = |S_{21}|^2 \quad (3.21)$$

Where S_{21} is the power received at port 2 relative to power input at port 1.

3.4 The proposed WPT System

Three systems have been designed and they are discussed in this section.

Most implantable medical devices have size limitations; for instance, a typical pacemaker has a size of 44×59×7.9 mm^3; hence, in order to realize a practical wireless charging system for IMD, both the receiver resonator coil and the load loop should be small, and the wireless power should be transferred efficiently from the transmitter coils to the small receiver coils with reasonable transfer distances.

To improve the transfer efficiency, [21] suggests the application of multi-turn coils for the receiver, whereas [11] uses the rectangular shaped receiver coil with a size of 10 cm x 5 cm. However, those designs are still too large to be applied to IMDs.

To derive a highly efficient and IMD-oriented WPT system, three designs were proposed and investigated here. They are shown in Figure 3.3, each of the proposed designs consists of four single-turn loops and no matching circuits are required. The geometric sizes and design parameters for each set of WPT system are listed in Table 1. All systems were designed to be operated at around 18.7 MHz. Initial designs were first simulated and analyzed with the High Frequency Structural Simulator (HFSS), and then fabricated. Due to intrinsic tolerance of the lumped capacitors, fine-tuning of the capacitors was performed to achieve the optimum transfer efficiency.

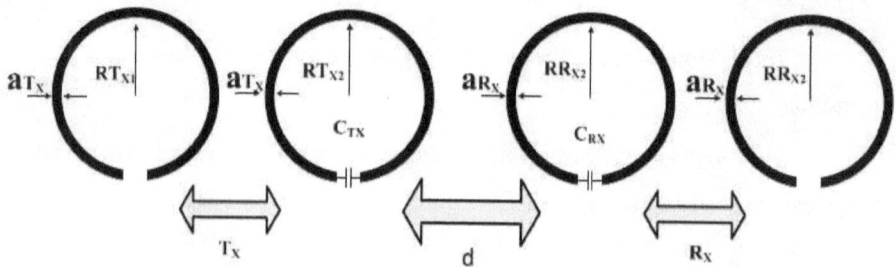

Figure 3.3 The proposed WPT system using magnetic resonance coupling.

Table 1: Parameters of Coils for proposed Three Designs

	Design #1	Design #2	Design #3
RTX1(mm)	150	50	50
RTX2(mm)	200	50	50
RRX1(mm)	50	50	25
RRX2(mm)	50	50	25
aTx (mm)	12.7	2.8	2.8
aRx (mm)	2.8	2.8	0.785
Tx (mm)	180	36	36
Rx (mm)	3.25	3.25	3.25
d (mm)	100	100	100
CTx (pF)	91	365	360
CRx (pF)	330	365	575

Chapter 4 Simulation, Fabrication and Testing setup of the Design

This chapter contains three main sections: simulations, fabrication and testing set up for all the three proposed designs.

In our proposed design, we have considered pacemaker as our exemplary IMD. The transmission distance of 8-10 cm was set as a benchmark in order to provide more freedom to the patients so that they can move around within the transmission distance range.

Simulation was carried out using HFSS (High Frequency Structural Simulator) [22] for all three designs. Fine tuning of each of the WPT designs was achieved through optimization both with simulations and by experiments. Efficiency is calculated by measuring scattering parameters $|S_{21}|^2$, $|S_{11}|^2$ and $|S_{22}|^2$ for each of the systems and than substituting it in equation 3.21. $|S_{21}|^2$ represents the ratio of output power to input power, $|S_{11}|^2$ represents the amount of reflected power back to port 1, $|S_{22}|^2$ represents the amount of reflected power back to port 2.

4.1 Simulation Results

4.1.1 Design #1

Here, the simulation results for design 1 using HFSS software is discussed. Simulation was carried out for the frequency range of 10 MHz to 25 MHz and the capacitor values were calculated using equation (3.20). The following graph shows all the scattering parameters at a transmission distance of 10 cm between the transmitter and receiver.

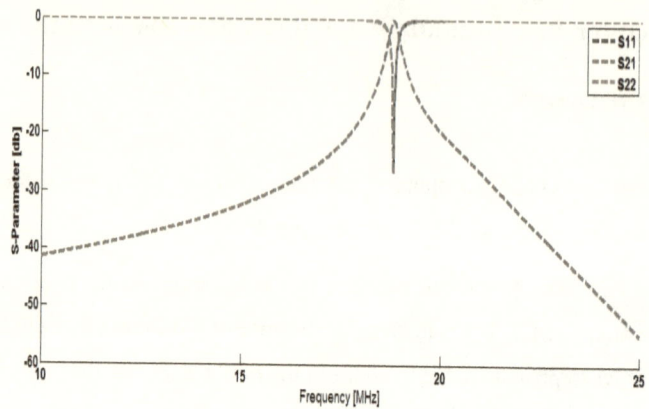

Figure 4.1 Simulated scattering parameters of Design #1

Figure 4.1 shows the simulated scattering parameters of Design #1. As noted from Figure 4.1, the simulation results yields -0.2309 db for $|S_{21}|^2$, -26.06 dB for $|S_{11}|^2$ and -26.8775 dB for $|S_{22}|^2$. The efficiency calculated with (3.21) is 94.67% efficiency at a 10 cm transmission distance at 18.78 MHz. This indicates that the system works on resonance as it has a strong peak at 18.78 MHz.

Several other simulations were carried out at different transmission distances of 5 cm, 10 cm, 15 cm, 20 cm and 25 cm respectively that yielded the efficiencies of 89.31%, 94.57%, 83.60%, 58.84% and 36.00% respectively as shown in the Figure 4.2.

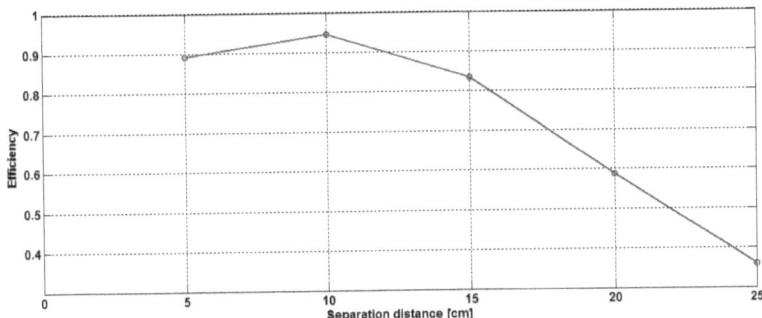

Figure 4.2 Effeciency vs transmission distance for Design #1

It is observed that even at less transmission distance of 5 cm the efficiency so achieved is less than that of 10 cm. This is because our proposed system works on magnetic resonance coupling and delivers the best results at specific transmission distance.

4.1.2 Design #2

Figure 4.3 shows the simulated scattering parameters of Design #2.

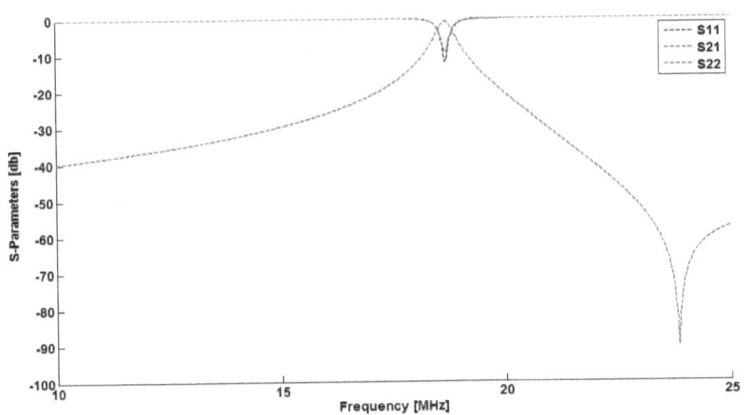

Figure 4.3 Simulation s-parameters for Design #2

Similar to Design #1, simulations were carried out with various transmission distances of 10, 15 and 20 cm respectively. Simulation results at the 10 cm transmission distance yielded $|S_{21}|^2$ of -0.94 dB, $|S_{11}|^2$ of -11.93dB and $|S_{22}|^2$ of -9.22 dB at 18.7 MHz. Efficiency at 10 cm is 80.53%.

4.1.3 Design #3

Design #3 is the final design of the present report. The size of the receiver is greatly reduced to meet the specification of the standard pacemaker. It is compact enough to be implanted in a human body.

Figure 4.4 shows the simulation results obtained for design #3.

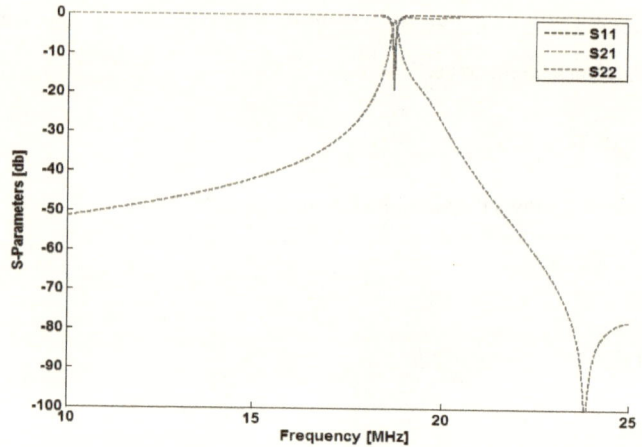

Figure 4.4 Simulated scattering parameters at 8 cm transmission distance for Design #3

As noted from Figure 4.4, the simulation results yielded -1.3235dB for $|S_{21}|^2$, -16.9885 dB for $|S_{11}|^2$ and -18.9975 dB for $|S_{22}|^2$. The calculated efficiency is 73.74% at 8 cm transmission distance operating at 18.7 MHz. It is clearly visible from the Figure 4.4 that the design works on resonance as it has a strong peak at our resonance frequency of 18.7 MHz.

Several other simulations were carried out at different transmission distances of 10 cm, 15 cm, and 20 cm respectively. The efficiencies achieved were 54.1% ,12.45%, and 2.80% respectively .

Once all the designs were simulated and satisfactory results were obtained, the next step was to fabricate the design system into hardware.

4.2 Prototype

Transmitter and receiver coils are built using a copper tube (alloy 122). Alloy 122 is flexible and does not break on bending.

The tubes were bent using specific tools in the Mechanical Department and were made to the specific dimensions used in the simulations. The stands holding the tubes were made up of wood. Optimization and fine tuning of the WPT system was achieved by adjusting appropriate values of the lumped capacitor soldered at two ends of the coil gap. Figures 4.5 and 4.6 shows the prototype of the coils for design #2 and design #3.

Figure 4.5 Receiver coils of 100 mm in diameter

Figure 4.6 Receiver coils of 50 mm in diameter

Once the hardware was made, the system was set up for measurements.

4.3 Testing Set- Up

During the testing, two transmitter coils were mounted on a wooden stand at the same height, and two receiver coils of design #2 and design #3 were mounted on each side of a 3.25 mm thick plastic sheet, the plastic sheet was then mounted on a wooden stand to maintain the alingment against transmitter coils. Test set up for the system is shown in Figure 4.7. Transmitter coils are fixed and receiver coils were moved by 5 cm, 8 cm, 10 cm, 15cm and 20 cm, respectively, for different measurements as shown in Figure 4.8. Distances amongst individual coils of both transmitter and receiver were maintained as per Table 1 parameters.

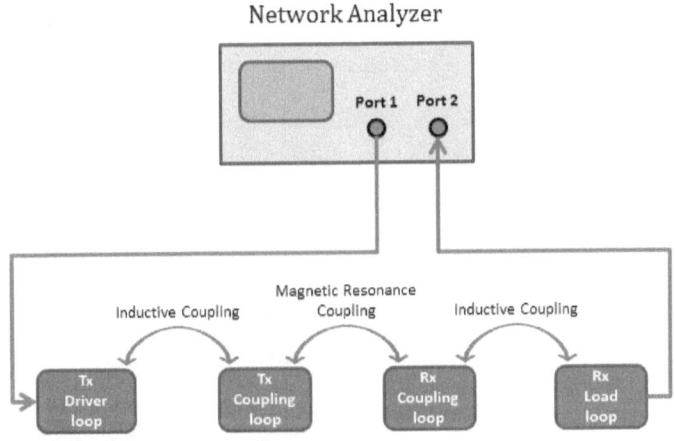

Figure 4.7 Test set up diagram

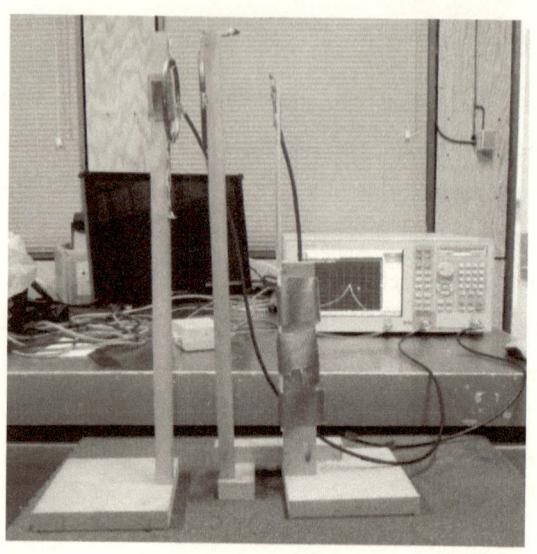

Figure 4.8 Experimental set up

The entire set up was placed on a wooden table about 100cm above a ground. The equipments used are Agilent E5071B Vector Network Analyzer (VNA) shown in Figure 4.9 and Agilent Vector signal generator (VSG) E4438c as shown in Figure 4.10.

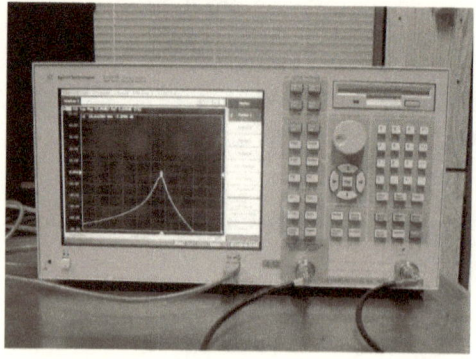

Fig 4.9 Agilent VNA(Vector Network Analyzer) E5071B

Figure 4.10 Agilent Vector Signal Generator E4438C

The practical prototype was designed and measurements were carried-out by maintaining the same distance as simulation between the two transmitter coils, the two receiver coils and between transmitter and receiver side. Fine tuning of the capacitors was achieved by using a high quality capacitor with less than or equal to 5% error tolerance. All four coils were placed in alignment and were attached accordingly to the wooden stand.

Chapter 5   Measurements and Discussions

In order to verify the design principles presented in Chapter 3, and to validate the simulation results presented in the last chapter, lab measurements were performed on the prototype. To provide an insight of the possible performance of proposed designs when implanted, the performance of the fabricated design #2 and design #3 under different working scenarios were evaluated.

5.1   Design #2

The VNA was callibrated before performing measurements, where the start and stop frequency were set to 10 MHz and 25 MHz. Signal was feed through the SMA (SurfaceMiniature version A) connected to the transmitter driver loop, through port 1 of VNA. The load loop at the receiver is connected through the SMA to the port 2 of the VNA. Measurement results at 10 cm transmission distance yielded $|S_{21}|^2$ of -1.32 dB, $|S_{11}|^2$ of -15.7dB and $|S_{22}|^2$ of -10.5 dB at 18.63 MHz as shown in Figure 5.1. There is a slight difference from the simulations in resonance frequency of measurement because of the tolerance of the capacitor.

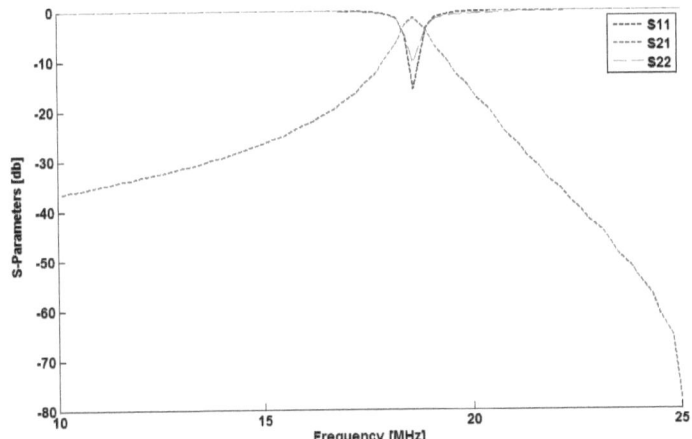

Figure 5.1 Measurement results for Design #2

Comparing Figure 4.3 and 5.1, simulation and measurement results for design #2 agree well with each other. Figure 5.2 shows the comparison chart of efficiency of the simualtion and measurement results vs various transmission distances for design #2. They collaborate very well.

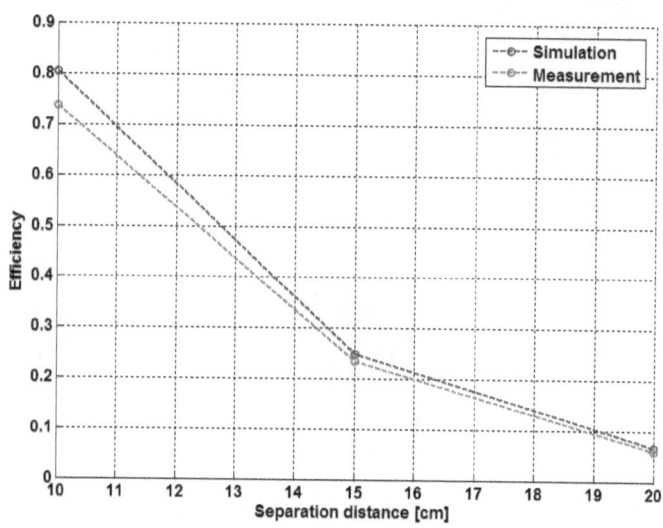

Figure 5.2 Efficiency vs distance for simulation and measurement for Design #2

Table 2 shows the efficiency in terms of percentage for simulation and measurement results obtained at various distances.

Table 2: Efficiency (simulation/measurement) vs Distance for Design #2

Separation Distance (cm)	Simulation Efficiency (%)	Measurement Efficiency (%)
10	80.53	73.8
15	24.84	23.38
20	6.61	6.00

The above results show the test gives almost the same results as the simulation and hence HFSS simulation is quite effective and accurate for our system design.

5.2 Design #3

Measurement results for design #3 are discussed in this part. Apart from standard measurements using VNA, measurements with pork skin and muscles were also verified by inserting the receiver coils with the load of an incandescent lamp into a 4cm width pork muscle. Furthermore, measurements at different angles were carried by moving receiver coils around the transmitter to see how effectively the system works when all the coils are not in alignment. The same procedure as design #2 was followed for the measurement by calibrating VNA and setting up frequencies and feeding the signal.

The system was feed with 2.36 volt signal from the vector signal generator. The receiver was then moved back and forth to see the distance at which maximum power was gained. The lamp glowed to its full intensity at a distance of 8 cm from the transmitter as seen in Figure 5.3 and gradually faded as moved closer and further away. The same work was carried out by inserting the receiver coils inside the pork muscle as seen in Figure 5.4 and still we were able to obtain the same results as one without pork.

Figure 5.3 Experimental set up without pork for Design #3

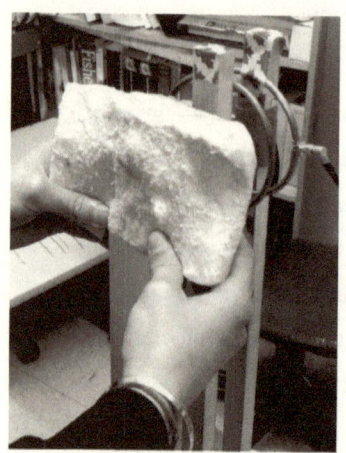

Figure 5.4 Experimental set up with pork for Design #3

Measurement results at the 8cm transmission distance yielded $|S_{21}|^2$ of -1.79 dB, $|S_{11}|^2$ of -18.9 dB and $|S_{22}|^2$ of -19.0 dB at 18.7MHz as seen in Figure 5.5, which in turn lead to efficiency of 66.22%.

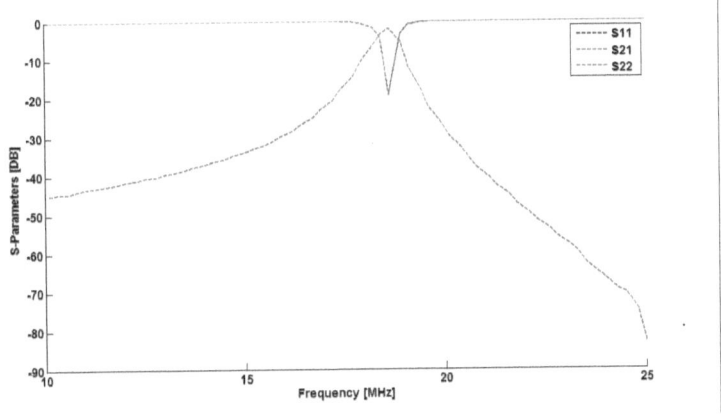

Figure 5.5 Measurement results at 8 cm transmission distance for Design #3

Similar to simulation, the measurements with and without pork were carried out at different transmission distances of 5, 10, 15, 20, 25 and 30 cm. Figure 5.6 shows the measured efficiency (with and without pork) at various transmission distances. It is clearly seen from Figure 5.6 that the maximum efficiency is achieved at transmission distance of 8 cm; whereas there is a decline in efficiency at the closer distance of 5 cm and also at the greater distances of 10 cm, 15cm and so on.

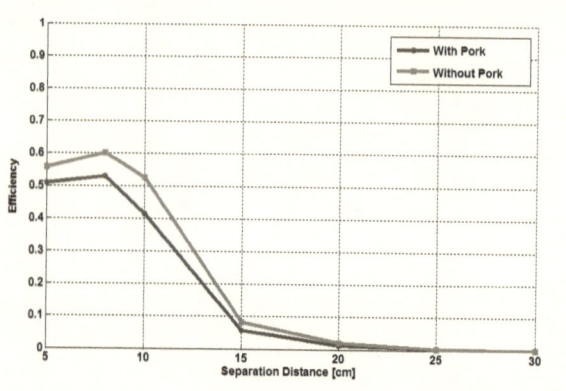

Figure 5.6 Efficiency vs Distance for measurement (with/without pork) for Design #3

Table 3 shows the efficiencies for simulation and measurement results (W/WO) pork obtained at various distances.

Table 3: Efficiency (Simulation/Measurement) vs transmission distance [with/without Pork] for Design #3

Separation Distance (cm)	Simulation Efficiency (%)	Measurement Efficiency(%) without pork	Measurement Efficiency(%) with pork
8	73.54	66.22	52.96
10	54.1	52.6	41.5
15	12.45	8.3	5.75
20	2.8	2.1	1.23

After achieving the desired results, the system was tested to exam its omni-directional capability. To our surprise, the system responded with similar results as the receiver coils are moved around at various angles from the centre of the transmitter coils. The alignment and the orientation angles for design #3 receiver coils are shown in Figure 5.7.

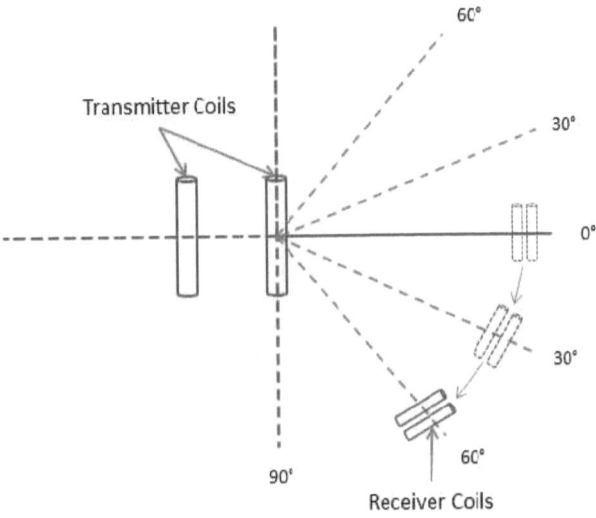

Figure 5.7 Orientation angles of receiver coils for Design #3

Figure 5.8 shows the efficieny at various orientation angles for the 8 cm transmission distance. The results obtained are convincing and the system efficiency does not degrade much even when at various angles. The need of all coils to be in alignment with each other is not required.

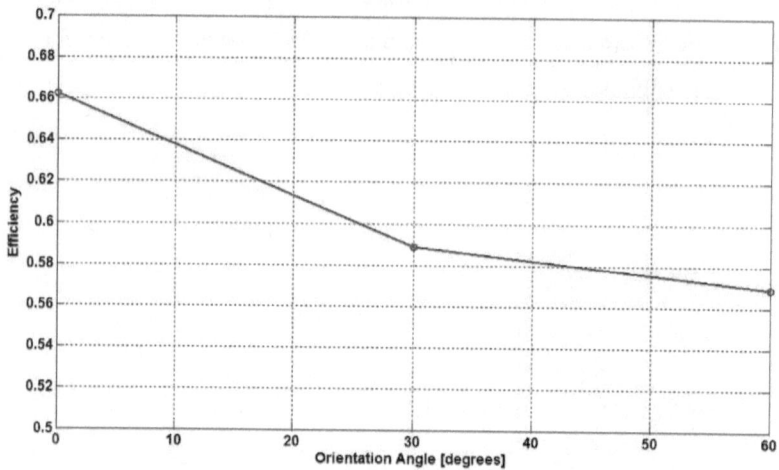

Figure 5.8 Efficiency vs Orientation angle for Design #3

In short, both simulation and experiment show that the proposed design is effective and efficient in wireless power transfer regardless of orientation angles.

In the following part, we have compared the performance of three proposed designs at 10 cm transmission distance in table 4. Table 5 highlights the advantage of final proposed design #3 compared to Fei Zhang design [12].

5.3 Performance Comparison of Three Proposed Designs at 10 cm Transmission Distance

Table 4: Scattering Parameters S21 at 10 cm Transmission Distance for Three proposed Designs

Proposed Designs	Simulation $|S_{21}|^2$ (db)	Simulation Efficiency (%)	Measurement $|S_{21}|^2$ (db)	Measurement Efficiency (%)
Design #1	-0.2309	94.67	-	-
Design #2	-0.94	80.53	-1.30	73.8
Design #3	-2.6684	54.1	-2.79	52.69

5.4 Comparison of Final Proposed Design #3 with Fei Zhang Design

Table 5: Comparison with Fei Zhanf Design [12]

Parameters	Fei Zhang Design	Proposed Design #3	Conclusion
RTX1 (mm) RTX2 (mm)	82.5 82.5(multiple turn)	50 50 (single turn)	40% size reduction per turn
RRX1 () RRX2 ()	20.5 20.5	25 25	18% increase in size
D (mm)	90	100	10 mm improvement
Efficiency (%)	22.35	52.61	30.38% more

As noted from table 5, the proposed design #3 transmitter coils are 40% smaller in size (single turn) as compared to Fei Zhang (multi turn) transmitter coil, thus reducing the over all area occupied. The size of the receiveer coils are increased by 18% in our design in order to match the cross-section dimensions of the commercially available pacemakers. The highlighting part of the design #3 is the improvement in transmission distance by 10 mm and efficiency by 30.38%.

Chapter 6   Conclusion and Future work

6.1 Conclusion

Three wireless power transfer systems using magnetic resonance coupling were designed, simulated, prototyped and tested. Good agreements between simulation results and measurement results were observed. With the radius of receiver coils reduced to 2.5 cm whereas the transfer efficiency retains higher than 50% in a relative large transfer distance range, the proposed design #3 can be readily applied to small implantable medical devices, such as pacemakers. Experimental results with biological objects validated the effectiveness of the proposed wireless power transfer system when implanted. Furthermore, the omni-directional characteristic of the proposed system renders it attractive for easy use of the system.

6.2 Future Work

Although, the system achieved the desired results as simulation; still some work has to be done as a part of future work.

- Investigation of SAR (Specific Absorption Rate) : how emitted energy from our system is absorbed by human body deserves investigation.
- Study of the possible heating effect if high power is transfered.
- Research of the impact of magnetic resonance coupling in near field RFID as interference.
- Other schemes to further reduce the sizes and increase the efficiency.

Bibliography

[1] T. Linlin, H. Xueling, L. Hui and H. Hui, "Study of Wireless Power Transfer System Through Strong Coupled Resonances," in *International Conference on Electrical and Control Engineering (ICECE)*, 25-27 June 2010.

[2] "WiTricity Technology: The Basics," WiTricity, [Online]. Available: http://www.witricity.com/pages/technology.html. [Accessed 24 January 2013].

[3] F. Hadley, "MIT news," 7 June 2007. [Online]. Available: http://web.mit.edu/newsoffice/2007/wireless-0607.html. [Accessed 27 January 2013].

[4] N. Tesla , "The Transmission of Electrical Energy without Wires as a means for Furthering Peace," in *Electric World and Engineer*, vol. 37, McGraw Hill, 1905, pp. 21-24.

[5] "Tesla Tower in Shoreham Long Island (1901 - 1917) meant to be the "World Wireless" Broadcasting system," [Online]. Available: http://www.zamandayolculuk.com/cetinbal/HTMLdosya1/Tesla-WorldWireless.htm. [Accessed 7 February 2013].

[6] W. C. Brown and E. E. Eves, "Beamed Microwave Power Transmission and its Application to space," *IEEE Transactions on Microwave Theory and Techniques*, vol. 40, no. 6, pp. 1239-1250, June 1992.

[7] "The Martian Chronicles," 20 June 2010. [Online]. Available: http://martianchronicles.wordpress.com/2010/06/20/our-burning-need-for-energ/. [Accessed 4 February 2013].

[8] "Space Technology Mission Directorate," NASA Government, [Online]. Available: http://www.nasa.gov/directorates/spacetech/centennial_challenges/after_challenge/lasermotive.html. [Accessed 18 January 2013].

[9] N. Tesla, "System of Electrical Lightning". US Patent 454,622, 23 June 1891.

[10] A. Kurs, A. Karalis, R. Moffat, J. D. Joannopoulos, P. Fisher and M. Solijacic, "Wireless Power Transfer via Strong Coupled Magnetic Resonances," *Science*, vol. 317, no. 83, pp. 83-85, 2007.

[11] J. Choi, J.-K. Cho and C. Seo, "Analysis on transmission efficiency of wireless energy transmission resonator based on magnetic resonance," in *Microwave Workshop Series on Innovative Wireless Power Transmission: Technologies, Systems, and Applications (IMWS), 2011 IEEE MTT-S International*, 12-13 May 2011.

[12] F. Zhang, S. A. Hackworth, X. Liu, H. Chen, R. J. Sclabassi and M. Sun, "Wireless energy transfer platform for medical sensors and implantable devices," in *Engineering in Medicine*

and Biology Society, 2009. EMBC 2009. Annual International Conference of the IEEE, 3-6 Sept.2009.

[13] D. Prutchi, "Implantable-Device," [Online]. Available: www.implantable-device.com/2013/02/18/second-sight-receives-fda-clearance-for-argus-ii-retinal-prosthesis/. [Accessed 16 February 2013].

[14] A. Johansson, "Wireless Communication with Medical Implants: Antennas and Propagation," Msc thesis, Lund University, June 2004.

[15] H. A. Haus and W. Huang, "Couple Modew Theory," *Proceedings of the IEEE,* vol. 79, no. 10, pp. 1505-1518, October 1991.

[16] H. A. Haus, Waves and Field in Optoelectronics, NJ: Prentice Hall Edition, 1984.

[17] A. Karalis , J. D. Joannopoulos and M. Solijacic, "Efficient wireless non-radiative mid-range energy transfer," *Annals of Physics,* vol. 323, no. 1, pp. 34-48, January 2008.

[18] R. D. Knight, Physics For Scientists and Engineers, 2 ed., CA: Pearson Education, 2008.

[19] A. Kurs, "Power Transfer Through Strong Coupled Resonances," M.Sc thesis, Massachussetts Institute Of Technology, 2007.

[20] C. M. Zierhofer and E. S. Hochmair, "Geometric approach for coupling enhancement of magnetically coupled coils," *IEEE Transactions on Biomedical Engineering,* vol. 43, no. 7, pp. 708-714, July 1996.

[21] B. L. Cannon, J. F. Hoburg, D. D. Stancil and S. C. Goldstein, "Magnetic Resonant Coupling As a Potential Means for Wireless Power Transfer to Multiple Small Receivers," *Power Electronics, IEEE Transactions on,* vol. 24, no. 7, pp. 1819-1825, July 2009.

[22] "Ansys HFSS," Ansys, [Online]. Available: http://www.symkom.pl/produkty/ansys_hfss.pdf. [Accessed 10 February 2013].

I want morebooks!

Buy your books fast and straightforward online - at one of the world's fastest growing online book stores! Environmentally sound due to Print-on-Demand technologies.

Buy your books online at
www.get-morebooks.com

Kaufen Sie Ihre Bücher schnell und unkompliziert online – auf einer der am schnellsten wachsenden Buchhandelsplattformen weltweit! Dank Print-On-Demand umwelt- und ressourcenschonend produziert.

Bücher schneller online kaufen
www.morebooks.de

OmniScriptum Marketing DEU GmbH
Heinrich-Böcking-Str. 6-8
D - 66121 Saarbrücken
Telefax: +49 681 93 81 567-9

info@omniscriptum.com
www.omniscriptum.com

www.ingramcontent.com/pod-product-compliance
Lightning Source LLC
Chambersburg PA
CBHW031548210526
45464CB00003B/1198